# How Long Can We Live?

A Book for the Forward-Thinking.

# Table Of Contents

How Long Can We Live? .................................................................................................... 1

Table Of Contents .............................................................................................................. 2

Chapter 1 – The Vision ...................................................................................................... 4

Chapter 2 – A Promising History And Future .................................................................... 6

Chapter 3 – The Quest For Answers ................................................................................. 7

Chapter 4 – Changes In Healthcare .................................................................................. 8

Chapter 5 – More Years To Enjoy ................................................................................... 10

Chapter 6 – An Everlasting Mind ..................................................................................... 11

Chapter 7 - Creation And Design .................................................................................... 12

Chapter 8 - The Body Is A Wonderland .......................................................................... 17

Chapter 9 – The Worldwide Issues ................................................................................. 23

Chapter 10 - The Artificial Fingernail .............................................................................. 26

Chapter 11 - Nanomedicine And Immunotherapy ......................................................... 28

Chapter 12 – Life After Death? ....................................................................................... 30

Chapter 13 – Innovations Galore .................................................................................... 32

Chapter 14 – A Youthful Attitude .................................................................................... 34

Chapter 15 – Hope And Nutrition ................................................................................... 37

Chapter 16 – Editing Embryos And Stabilizers .............................................................. 43

Chapter 17 – The Artificial Kidney .................................................................................. 46

Chapter 18 – Stem Cells .................................................................................................. 49

Chapter 19 – Cell Regeneration ...................................................................................... 51

Chapter 20 – Mental Clarity And Direction .................................................................... 52

Preface

This book is dedicated to Nick, my wonderful love, may we love each other forever!

© 2017 Bethany Healy

## Chapter 1 – The Vision

I originally was thinking of writing this book instead as a patent for my drawings of human body part designs, but then became inspired to have a book for all those who are interested to read about the topic of ever-lasting life and human longevity. What if we lived in the future where there's a possibility of people lining up for anti-aging procedures such as prescribed medicine for cancer prevention – like a vaccination, and where people are concerned about looking younger for longer than ever before? What about a world where people lived to be well past 100 like some people are already doing today, and what if we all were thought of as normal for living this long? What if we became aware of how to prevent such diseases such as cancer, heart disease and diabetes? And we could change our children's DNA by editing their embryos before conception? And, what if people became interested in artificial body parts, including memory implants in the brain and better heart valves and organ functions that could make us live even past who knows? 200 years old? These are some real possibilities to some very educated individuals. And, I happen to have a passion for envisioning human prosthetics that are very realistic and functional for every body part on the human body, including organs, bones, hair and fingernails.

Most Christians believe the soul of the body has ever lasting life but the body does not. Well, obviously, the body does not – right now. But there are millionaires out there investing in prolonging the body and it is believed by rich entrepreneur Dmitry Itskov that we can somehow transmit memories and life from the brain into an avatar for continued life after the death of the body and he's working on a project projected to be finished by 2045. I think that's far out there, but hey we used to believe televisions were impossible until everyone owned at least one! I think there are too many precious things about the body that we need to take into consideration keeping our bodies as strong, true and in-tact as possible by using replacement body parts along with being healthy. I have designed prosthetics using drawings and ideas, some of which I will share throughout this book, including tips on how to do it yourself.

Prolonging life has been a topic I've been thinking most about since the passing of my beloved grandmother from my mother's side and grandfather from my father's side, both of whom died within a short while of each other around the age of 70 of terminal cancer. They were like second parents to me. It was very difficult to watch them suffer and die. They had a nice long life by today's standards. They had very rich family lives and were happy people.

I started thinking of ways we could live longer, and I wondered why doctors didn't try harder to save them. I know they did their best, but it just wasn't good enough for me. With all the standard protocols, they and nurses were following, it seemed they were just shoving poison into my grandparents' bodies and hoping for a cure. I know we have yet to develop drugs that work on cancer for certain, but it was awful. And, they just ate regular food. I was feeding my grandmother mashed potatoes in the medical care center. I felt like they should have been on a special cancer diet with green shakes that included spirulina, broccoli and kale with special fats like from avocados and such and they were instead under nourished and deathly skinny. Our health systems need a big change for the better with more nutritionists and anti-aging specialists leading the way.

## Chapter 2 – A Promising History And Future

People have consistently and gradually been living longer over the years. Not long ago we used to think it was normal to die around age 50 and today the average lifespan of women is 80 and men is 75. People from back in the day would think you were crazy to think we could live to be 100 or more, but there are more centenarians now than ever before! I hope I'm not thought to be crazy for shooting for 200, because I feel that doubling our lifespan would greatly increase the longevity and quality of life without going overboard and becoming robots. And we doubled our lifespan in the past 100 years or so. I'm not just dreaming of an everlasting life with no results. I think 200 gives us a good way to shoot for the moon because even if you miss you'll be among the stars. I'm aware that this may not happen for me or anyone I know. I'm simply being speculative and hopeful. It is surprising how much people are interested in living longer, yet how little is currently being done to research preventing aging comparatively speaking. Most people just accept aging as a natural process, but it is anything but natural. People die of all different causes at various times throughout their lives and very little answers are out there, but some people are getting closer to developing some real answers. We must expect more. We would never expect to go back and think that 35 was an acceptable age to live to anymore, so why then should we not expect to overcome obstacles in our way now that could make us live longer and longer? That's the goal, right? Researcher Aubrey de Grey believes the first person who will make it to 1,000 years old is already alive. He founded a non-profit organization dedicated to extending the human lifespan, and was given 6 million dollars from rich entrepreneur, Peter Thiel.

# Chapter 3 – The Quest For Answers

Don't get me wrong - we're doing great! With today's medical advancements we are really getting somewhere with all the new medicines and technology. I think it is wonderful that we are climbing a steady hill of success along the lines of living longer. At least we're not going backwards and trying to fend off some sort of disease like malaria - we've already conquered that! Some people think of aging as a disease by the way. And that there will be a cure. Can you imagine what it will be like when it will be normal to get a memory implant in the brain and get skin treatments that actually prevent wrinkles, and other age defying and preventing procedures like these as standard protocol? The world will be much different. Scientists and doctors are doing as much as they can in effort to find the hidden answers to living longer. Stem cell research has been going on for decades and I have an uncle who saved his now 12-year-old daughter's umbilical cord for stem cell purposes. People are doing things like this! Scientists are learning how to grow human organs with stem cells so that we may have real replacement body parts. And they can grow parts of parts, so that tissues can be rebuilt over time. This is one of the most promising studies for the future of regeneration.

I started making designs for robotics that could be used in the human body to transform lives into working properly again. An example of this would be the pacemaker, which I obviously did not design. But I wanted to use it as an example of robotics already working in the human body and show there is a possibility of these designs really working. Our parts are inevitably going to wear out as time goes by, and we need to be able to replace them with either real or artificial body parts. And, we have the technology to conquer any body part, which we are already starting to do.

# Chapter 4 – Changes In Healthcare

I believe at some point General healthcare will be available to all for free or at the fraction of the cost to promote the growth of medicine and robotics for health and human longevity. Eventually we'll all see the benefit of Universal Healthcare, because it should be available to all Americans and the rest of the world to a certain level. We are the only advanced nation without it now. There are currently free clinics and government assisted programs like Medicare and Medicaid to help provide for people's health concerns and problems, but everyone needs free or low-cost health care. I think eventually in 10-20 years or so we will conclude that while healthcare in the private sector is great, healthcare for everyone is much more beneficial to the discovery of new health measures that could be used to save more lives. And, the more people demand it the closer we'll get to having it. I think there will still be room for private sector healthcare but that it will be for more specialized care, such as anti-aging and preventative procedures, along with special designer prosthetics. Doctors are already coming up with ways to help those with a likelihood of getting cancer by giving them preventative medicines. Soon, it may be normal for people to get a cancer vaccine.

As for now, the costs of healthcare are enormous and people are spending tons of money on anti-aging plastic surgeries and other procedures that aren't covered by health insurance. People want to look younger. People want to be healthier and the cost of specialty gym memberships is sky high, at $35 per class in some areas. Healthcare will take into consideration people's efforts to take better care of themselves in the future and those doing more work will be rewarded by not being obese and by maintaining a healthier outlook on their lives. I'm not saying people who get plastic surgery will have cheaper healthcare, but I think those concerned with prolonging their life and living a healthier lifestyle will be looked at as less of a risk to healthcare providers. Rewards will be there for people being proactive about their overall health and outlook on life. When you look good you feel good, and some people think they look better with certain procedures that minimize under eye puffiness, nose problems, wrinkles, double chins and other unsightly issues. This may be viewed as a healthy way of thinking to have a better mental state of mind and prevent depression and anxiety and improve happiness. There will be more ways for these types of procedures to be possible for more people. Costs will go down. And more treatments will be available to more people. This is just how it works in our monetary society – older things generally get cheaper. Plus, there is a thing called Care Credit, which is like a credit card that allows people to get these procedures and make payments on them. It will become easier to get specialty procedures, and specialty procedures are where the future lies.

# Chapter 5 – More Years To Enjoy

So, what will we do with all these longer living, different and better-looking people? What about the people who have artificial limbs and have a beautiful prosthetic that is very functional? They are part of the future of human longevity and betterment too. I think there will be an emphasis on family planning and more people will have less kids or no kids and life quality will improve. We will have to grow more educated about population control or all be living in tiny houses with more growth of the number of people alive, which is growing exponentially. We will have to watch out carefully for our carbon footprint to prevent too much global warming and pollution. It will be important to focus on quality not quantity. I think less people will choose to have kids and more people will choose to have fewer kids. There will be a closer bond for those alive since they will have more years to spend together. And, people may be more loving and kind in the future when they know it matters more, because we will all be around longer. The stakes will be higher. I believe people will get better educations.

More and more states will be offering free college educations and the information available to anyone will be even better than it is today. Education is the leading way we can learn to have world peace and prosperity, along with learning how to all get along with each other and work together to form new innovations and futuristic products.

There are plenty of researchers thought of as crazy – especially at first when they think and speak of doing their work, but really that way of thinking is what makes someone a visionary and a pioneer. Never let anyone tell you you're too out there for being a creative genius. We need more people like that in the world. And, with creativity being a better education development we'll get there.

# Chapter 6 – An Everlasting Mind

There are neuroscientists out there now who are putting silicone implants in rats, monkeys and human brains – the hippocampus - that help with creating the possibility of establishing long term memories through electronic devices that measure electric impulses in the brain related to the creation of memories. Patients with severe memory loss are starting to be rehabilitated with these implants, and they've had some success. One of the geniuses behind the work is Theodore Berger, a bio medical engineer and neuroscientist at the University of Southern California in Los Angeles and Harvard graduate. He's hoping to restore the ability to create memories by implanting chips like these in the brain more often. 12 volunteers with epilepsy had these implants implanted in their hippocampus. The algorithm created an 80% correct memory recollection when shown pictures and asked to recall them 90 seconds later, compared to the brain's actual firing patterns. It's not perfect yet, but they've got a great start according to Berger! It may sound far-fetched, but it's true, and neuroprosthetics are only the beginning of our ability as humans to create the technology behind human longevity of life.

The brain is by far the most complex organ in the human body and we're already achieving leaps and bounds of success. It's only a matter of time before we're choosing designer prosthetics from a catalog to be installed in my predictions. It's exciting stuff! It's scary to some people, because they imagine privacy issues and government involvement as well as just the fear of cost and surgeries in the future, but to me I couldn't be happier to discover and possibly help create the advancements in technology. Yes, we may be dependent on machines, but so what! At least they can be repaired and replaced.

## Chapter 7 - Creation And Design

In general, the thing I've noticed that helps with creativity in designing parts for the body that could work as say prosthetic kidneys or an artificial heart valve is to increase the size of the model you are working with. So, let's say you're working with the kidney, blowing up the image on the page so that all your machinery can fit into the design is how to really get detailed. Anything is possible with nanotechnology. People can create prototypes from these designs.

So, by blowing up the picture it increases the ability to develop nanotechnology, and more people can begin developing designs that could work for helping with robotics and even medicine. I created models of the heart, skin, liver, kidney and entire body that can be used to configure drawings of artificial body parts that not only make it more comfortable for the person, but also increase the likelihood that people making the products can create something functional.

I studied Business in college online and I wrote about 3M in one of my essays. I think we need to develop more companies like 3M that can provide good things like quality band aids and colostomy bag hole covers and such to help develop some more of these ideas. Better colostomy bags are on the way too. We need to develop products that are not only comfortable but aesthetically pleasing, such as a designer colostomy bag that one cannot see through. Creating comfort for the patient is important. The more companies out there creating innovative products for the human body and medical world, the better off we will all be when it comes to having surgery, getting older and having changes happen to us that create the need for products like artificial organs, second skins, and even hair transplants and artificially installed hair.

When we work together, it's simply better. The organization of all these great ideas is also going to be key in creating a functional system for growth. In other words, let the ideas pour out - don't worry too much about the details and where the money is going to come from. That will all be there and that is what books are for. Copyright protection is wonderful! I want my ideas to spread and help others in the process. That's how we're going to get somewhere! Like Nike says, "Just Do It." If you have the money for patent attorneys to help you protect any new ideas you have that is even better, but not completely necessary to protect your ideas.

I've always had a very creative mind – I started writing songs when I was just a child; eleven-minute piano songs that I could remember how to play exactly! I was considered a child prodigy and even won a national songwriting contest by age 14 where my mother and I were flown to the Rock'N'Roll Hall Of Fame where I performed my original music and made a Levi Strauss-sponsored commercial that aired in Levi Strauss stores nationally and was on Chanel One – a network seen by 7 million students at the time. I was also on several news channels. Creativity is in my nature. I love art as well, and I've painted several paintings.

On to the meat and potatoes of it all. I foresee a future of the human lifespan mainly being prolonged using robotics. We have the materials and we have the know- how and the equipment is all there. We can do it! We already are, so it is just a matter of growing and improving what is already done, along with coming up with new improved processes and models. I'm not suggesting everyone is capable or even has the desire to create prosthetics, but for those who do or who are interested in learning about it, I am sharing exciting information.

I'll include some examples of drawings throughout the book so you can clearly grasp what I'm speaking of when I say blow up the size of the picture to a working level. I once imagined a wall of just a skin sample so there would be no mistake in the creation of artificial skin should anyone need it. Think Allison Wonderland. Making things large and small again. I believe we can create all seven layers of the skin and we are already performing skin grafts and the like. I'm specifically talking about something wearable that can be used as secondary skin so that as we age we can pick out new skin if we choose to do so. It will be like choosing a new outfit to wear. I am getting a little ahead of myself, but you get the idea – futuristic thinking is needed if we're going to make strides.

These design practices can be used to create anything really. Dentists can use them to create better products for their patients. And I think design is the wave of the future - everyone's going to be expressing themselves and doing it! Don't worry about patents. Just design and get the ideas out there. Worry about the legal stuff later. You can copyright drawings. I choose to use copyrights for my artwork, songs and books for my ideas because I find it the much more efficient and less expensive way to do things. So here, for example is a hair follicle; the larger the better:

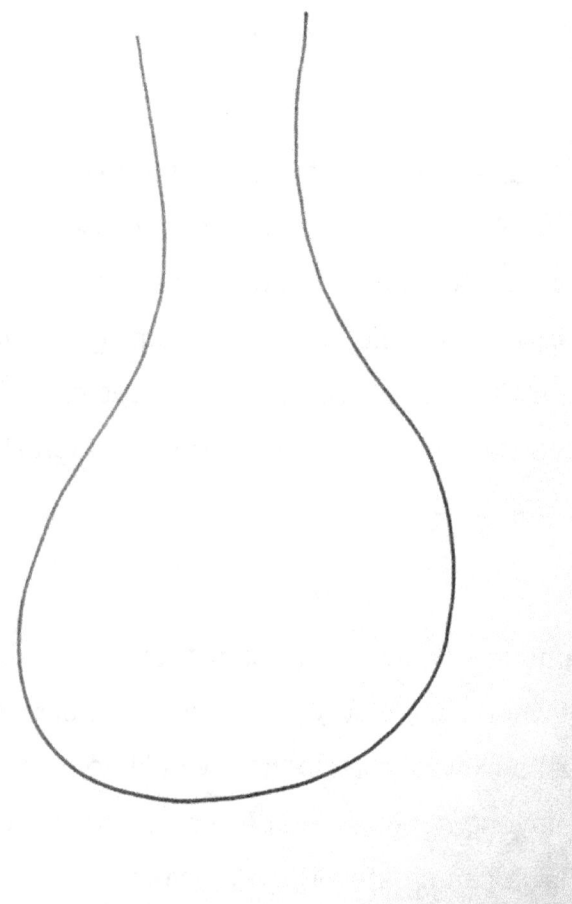

Versus:

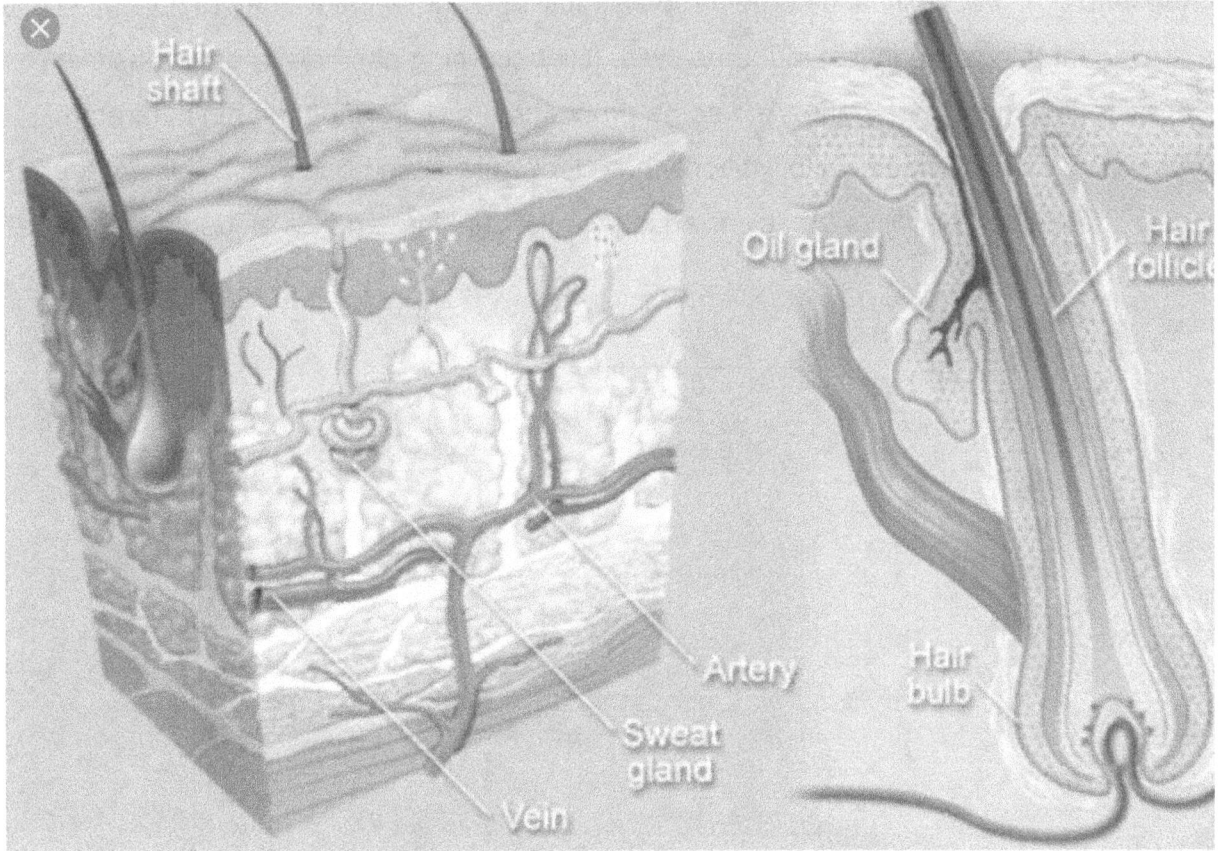

There's always time later-on to add the extra details, but don't let them get in the way. You can really dig in and do the dirty work of design when the image is large enough, and you can always add more details later. The thought of manipulating the body may be scary at first, but we cannot be afraid of manipulating a fake body. Think of a model or plastic body that is made with rubber. When you think of design think of creating a body (or body part) that isn't real instead of working with a real human if it helps create. The function is the main goal. I like prosthetics to be realistic – realistic looking skin and the like, but these days anything goes. People are piercing their bodies and manipulating their skin with tattoos and earrings that stretch out the skin of the earlobes. This is the future! We are already thinking along the paths of design and creativity of the human body. Some like to take it further than others. People are getting plastic surgery more than ever. It's really a personal choice and financial decision. Some people choose to *age gracefully* and some people simply don't have the money to do extravagant things like invest in themselves. It is becoming easier with programs like Care Credit to get things done. And some procedures are more popular in different areas such as Los Angeles, Hollywood where looks are a big concern. It's all a matter of taste. There's something to be said for each way of thinking, and it's all personal choice.

## Chapter 8 - The Body Is A Wonderland

When I see a blank page of paper I see endless possibilities. We can create! I don't have all the answers, but I believe all the answers will come to everyone working on longevity. I personally don't believe we can live forever yet, but I do believe we can live longer and that is a fact based on history. So, with all these facts in place it's only a matter of time before remedies are put into action! I meant for this book to be very readable, like "Who Moved My Cheese?" with very few pages and simple concepts. I hope you're finding it to be a good read.

So, what does a replacement kidney look like? I think it should be the real thing. There are already many designs. They have models out there, but I think there should be *parts of parts*, and that they should be able to be installed rather effortlessly and the person out within a day of surgery

So, my question is what do we need? And we should all be asking this question, but as a designer of prosthetics; what is it that we need exactly? Someone out there needs something in or on the body. There are so many people in the world and everyone has a unique need. I see a future where we're all excited to pick out a new moveable, realistic leg if we need one. And there are designer choices. This goes for every body part! We need to think this way to prepare for the future. If we're going to create a needle that can distribute medicine and vitamins to a hair follicle to help the hair regrow or grow back stronger, or disappear for that matter, then, we need to have the right gear. We have the technology to create this needle. It's similar to acupuncture.

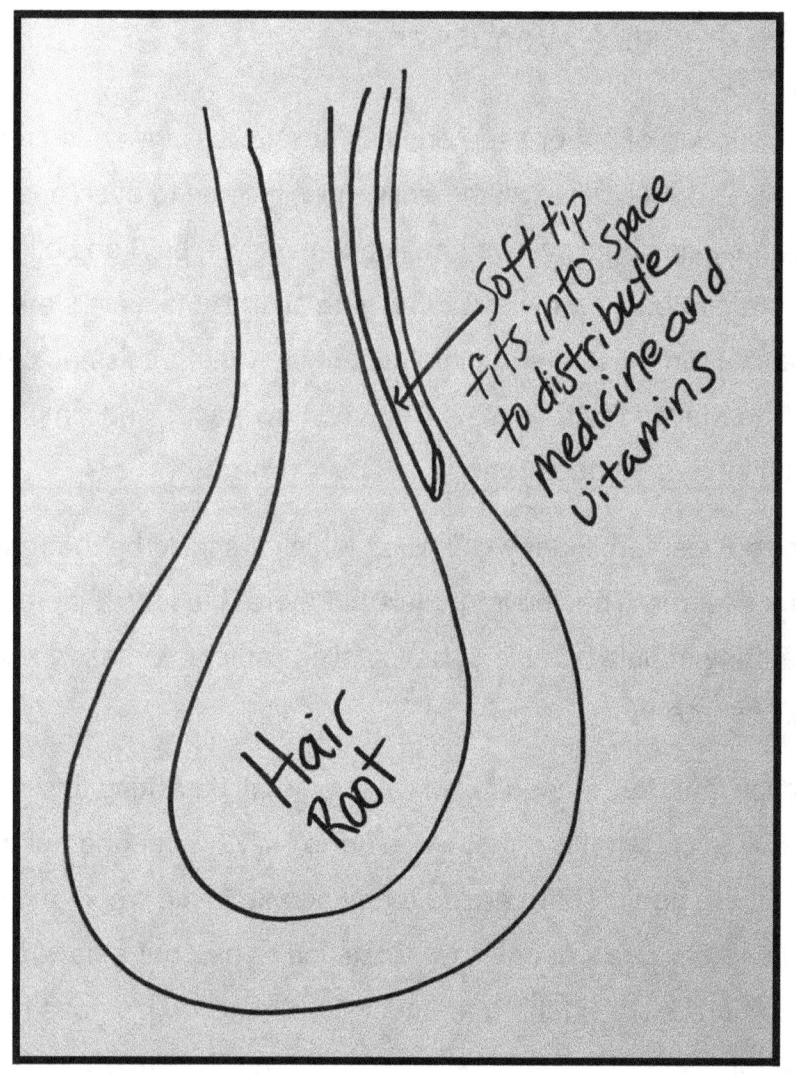

Scientists are discovering how to create implanted electrodes in paraplegics to move robotic arms with their thoughts. They are helping them learn new technology to live better lives. That means our thoughts are becoming less of a mystery to brain surgeons and we are becoming capable of restoring brain functions that were unable to do in the past. Where is this all going? What will we do with a longer and better life? Enjoy it! How will it affect relationships? Hopefully they will become even more meaningful than they already are. I'd like to see more art festivals and interesting museums showing the possibilities of new prosthetics and plans for longevity of life. I've often wondered why more elderly don't replace their frail limbs with stronger prosthetic limbs. Designs for prosthetic body parts are very possible and very realistic. They can be done for every body part using nano-technology and soon we'll have better 3-D printers that will be able to print them out. Perhaps there will be a popular age to get prosthetic limb surgery in the future to prevent aches and pains and ease the transition to older age. I know one major sign someone is getting older is their back curves. Getting chiropractic care early on is crucial in developing a healthy and strong spine throughout our *entire* lives. I used to get migraines until I had my neck adjusted and it has been over 10 years without a migraine. Chiropractors are on the cutting edge of healthy aging.

I hope more rich millionaires choose to invest in the medical field to help their own lives last longer if for no other reason. We need more support of those who are studying ways to increase longevity and we need more people studying and researching ways to live longer, which is why I'm encouraging design. So many people hang on to their money and do not invest it. But if more people who could invested in these researchers on the cutting edge of technology we would get faster results and better results. I think the more designers we have the more options we have as far as creating what we need exactly to make the progress happen, which is why I love designing so much.

Another key to design success is simplified drawings. Instead of worrying about all the details of the drawing and graphics, do something streamlined to get the creativity out. Such as, instead of drawing a human head, draw a question mark-like image. Then it's easier to design.

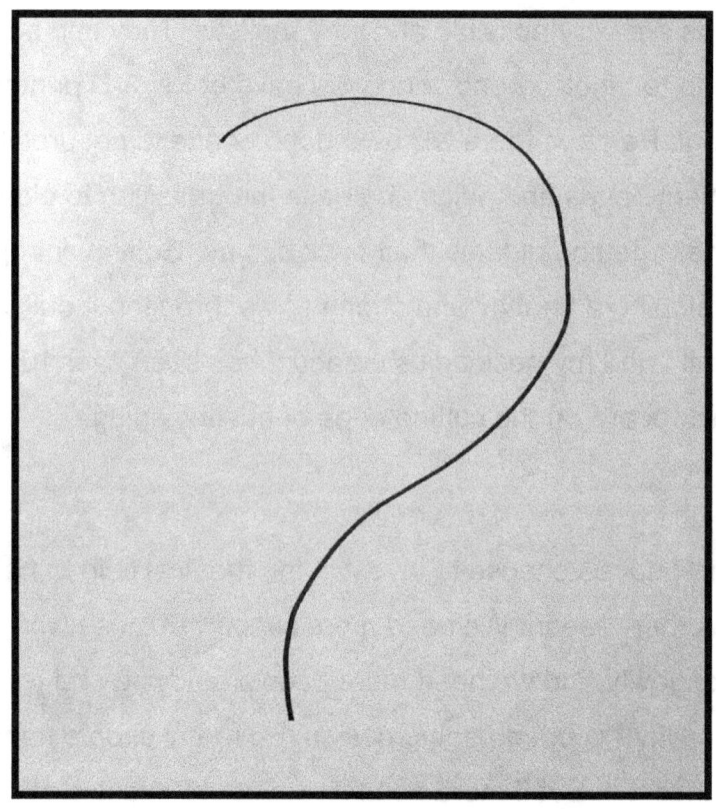

I have drawn a ton of closures and clasps that could be used in attaching prosthetic organs to the interior of the body. The bones are structure points to which stabilizers could be attached to create a stronger body as we age. My great grandmother who has passed on recently, was in her 90's and she had wonderful working hands that she used to knit gifts for little starving African children so they had dolls to play with up until her last years. She was amazing and in her care center she made tons of crafts and had a lot of fun doing it. She had a very curved back, which is why I would highly recommend getting appropriate chiropractic adjustments for at least a little while in one's early life to straighten out any scoliosis and abnormal curves in the spine. Having good posture is very important too, because it helps the body function better inside and keeps the spine in alignment. It also has an energizing effect. She also struggled with prolapsed bowels in her latest years and while that is not a topic I easily write about, it needs to be addressed. I designed some slings that could be attached to the rib cage and hips to help hold up the colon and strengthen the elimination system in our bodies. These would need to be installed at a younger age and with futuristic surgery procedures that aren't yet available today. Although, there is a procedure whereby the colon is lifted, but it is relatively new.

There must be ways of re-creating every function in the human body. I think there are multiple ways to do everything and it is just a matter of creativity. I love to paint and be very creative and I can think of countless ways to latch something together. It is a challenge to latch together some slippery body parts to do something like a kidney transplant. But, there are plenty of grippers and grippy material that can be helpful. Clamps and the design of several kinds of clamps could be helpful, along with appropriate stitching.

I designed eyelashes that could be sewn into the eyelid using micro-technology. They would be a permanent solution for those wanting longer eyelashes. Although, there are new technologies with lash-growing serum that help the lashes grow longer, naturally. Some people just don't have eyelashes because of a disease or hereditary and they may be interested in the false eyelashes that are permanent. Also, some people just want longer thicker lashes. I think a plastic lash would look great on anyone. They can be made to look perfect and they would be relatively easy to install under the lash area skin. Also, I think if someone want to get plastic surgery they should. Life is short and you need to feel great about yourself. I think it's worth it if it isn't overdone. People can get addicted to plastic surgery creating a face that looks like playdough. And that is a very disturbing look. But a natural touch up here or there can't hurt. It is very costly, though. Some people are more concerned with looks than others, and that is okay. To each their own. We can all get along!

Designing can be enhanced with an increase in numbers of designs as well. For example, there are countless ways of doing things. Coming up with 1,000 ways to attach something regardless of how well it will work could be beneficial in finding that perfect way to attach something. I've designed a page full of small latches and really the possibilities are endless when it comes to design. Human expression is a wonderful part of life and it should be celebrated.

# Chapter 9 – The Worldwide Issues

Things getting out of control and human growth together is why leadership is so important. I think if all world leaders got together they would be able to influence and learn from each other, as they do now, but better. Like with educational seminars at the events. And education will be even more important! I think people will be getting multiple degrees and education will be free or very affordable in the future. Strong leadership with a developed focus on individualism will help make our future as a human society strong. The more we learn to live with our differences and accept each other's different viewpoints, the better off we'll be in creating a sense of freedom for all. I think as the West influences more of the world, people will become more open-minded and free thinking.

What does this say about the inequality gap regarding finances? Would more poor people die off from not living as long while the rich just live longer and longer? I hope to see government programs in place that eventually make it possible for everyone to get preventative care and not just see it as anti-aging, but as a real medical necessity. Perhaps at first only the wealthiest will be able to afford certain procedures, but then prices will lower and it will be available to all.

What about the creation of mass destructive weapons? Life will be even more precious and we will be even more concerned as loving people to keep ourselves safe. What are we going to do with all this creativity? Hopefully not create more mass destructive weapons! That is a scary thought. But, I think we'll have programs in place to help eliminate the risk of faulty use of weapons on that level. It is inevitable that the creation of these types of weapons will multiply, but with the development of processes guarding who gets to use them I think there will be standard safety protocol that will be strictly followed.

What's the world going to look like outside? Are we going to be able to recognize each other? Will we look like space cadets? What will our fashion be like in the future? What are people going to look like? What are people going to listen to on the radio? We can't predict the future, but we can rest assured it will be different. So why not start thinking differently today? At least having open minds is necessary for the acceptance of new possibly brilliant ideas. Thinking futuristic thoughts is not just for neuroscientists.

I see and hope for a world where implants are put into the brain that can track brain waves and help us create memories and keep memories in a photographic and videographic way. Maybe curing Alzheimer's disease. This I have no certain idea how can be done, but I believe it can be done based on our technological capabilities!

So how will privacy and government be affected? I think people will want their privacy to be protected and it will be. I think things will develop naturally in due time and everything will fall into place. I don't think there will be a revolution like with The Pill where everyone started making love with one another. These concepts will be fairly easy to grasp and people will want to be involved. There's nothing scary about it! To me it's very exciting! And it's nice to think living day to day and enjoying every moment is the way to go - it is! But being proactive about health is all I'm really saying that's important. Too many people are overweight today. Too many people die of diabetes and heart disease. Adding shakes with greens in them to your diet can't hurt. Some people choose to have a paleo diet, which is an extravagant diet based on many principles including eating special fats including saturated fat and coconut and avocado oils, etc. But I prefer to not think too hard, since many people living to be above 90 ate bacon fat daily and white processed sugar and a lot of it must have to do with family history and genes.

People can already create micro machines that do incredible things. Think of the factories that have tiny robotics to be able to create things. These can be used in the medical field. All I'm saying is that anything can be done. We have the intelligence to do the things we need to do. In my opinion, there is no way we can jump to 110, 120…200 years of life without implementing robotics into the picture. Parts wear out – we will need new ones. And, with today's technology- it's possible. Plus, with stem cell research, the parts may even be able to be real.

I will say that I think people should try harder to make relationships work these days. Where will marriage go if we're going to be able to live so much longer? Relationships with age gaps will be more normal since we will have so much more time. Maybe some people will want to stay together because they'll have even deeper, more meaningful relationships and maybe some people will play the field more because they have more time. I think we'll see a variety of viewpoints and there will be an increase in individualism.

What about all these nations that are having so many children? Are they ever going to become educated enough to realize quality of life for the kids improves (and the parents) if there's more family planning? I'm not against big families – I grew up Catholic and my only brother has five children that I adore. I am not planning on having any kids, because my boyfriend already has two kids and I don't need to. Dolly Parton didn't have kids. Some people need to realize that there's a lot to think about when raising a family and not everyone needs a big family. The cost of raising a child is tremendous and now with lives lasting so much longer I think the focus should be again on quality, not quantity.

# Chapter 10 - The Artificial Fingernail

Here's a freaky thought for you… our hands are usually bony and frail in today's old age. What if our fingernails are taking the nutrients away from our body? The nails seem to get stronger as we age. What if there's a day when we need to replace our nails with artificial nails? Not just as fake nails, but real nails made of plastic that are implanted into our finger bone and replaced with a nail décor that can snap on and off to have beautiful interchangeable nails that don't take away nutrients. I have no proof that this is true; I'm merely speculating. But, what if it were true? Where would we get this nail surgery? At a nail salon with a numbing needle? At a hospital? I think it may be at a specialty surgery center, and that may be where health care is headed. The specialty centers are going to be the costly places where insurance is covering people, and normal, routine healthcare will be free, or government paid. We could buy snap on nails at Walmart instead of glue-on nails or paint. Would this improve our health? I don't know. But it sounds fun to me. I'm not afraid of surgery though.

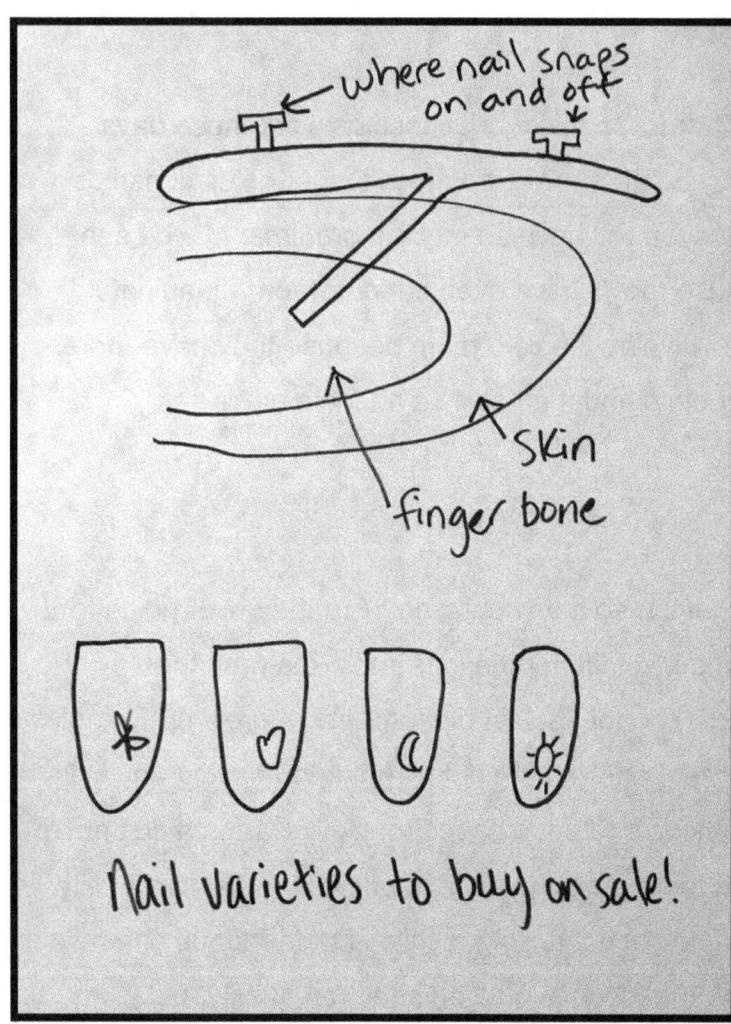

All I'm saying is there are plenty of possibilities for the future. We have the technology to make these things happen. We just need development and people may soon be getting fingernail implants. Maybe there will even be a guitar pick fingernail! No more painful hangnails! No more annoying polish changes or chips! It could be a possibility.

Where do the lines of robot and human blur? We're already creating shows like "Humans" and "West World" where robotic-like humans are existing. Is that the future? Where are we going with all this? Technology can be scary and mind-blowing at the same time. I love the thrill of it! Can we live forever? Perhaps in time we will find a way to replace all our body parts and be able to live forever. But at this point I'd be ecstatic with 200 years. Doubling our lifespan would add so much. So, while 200 years is a long shot, I think it is something to shoot for to keep big progress coming.

# Chapter 11 - Nanomedicine And Immunotherapy

Nanomedicine is also a new way designing can benefit the human body to increase longevity. What happens is tiny of the tiniest particles are implanted in the body and they help distribute medicine evenly, over a long period of time helping the body absorb medicine and target specific conditions in a way that regular medicine cannot do. Here's a model that shows some different types of nano technology and the relative size. A human hair is approximately 80,000 nanometers.

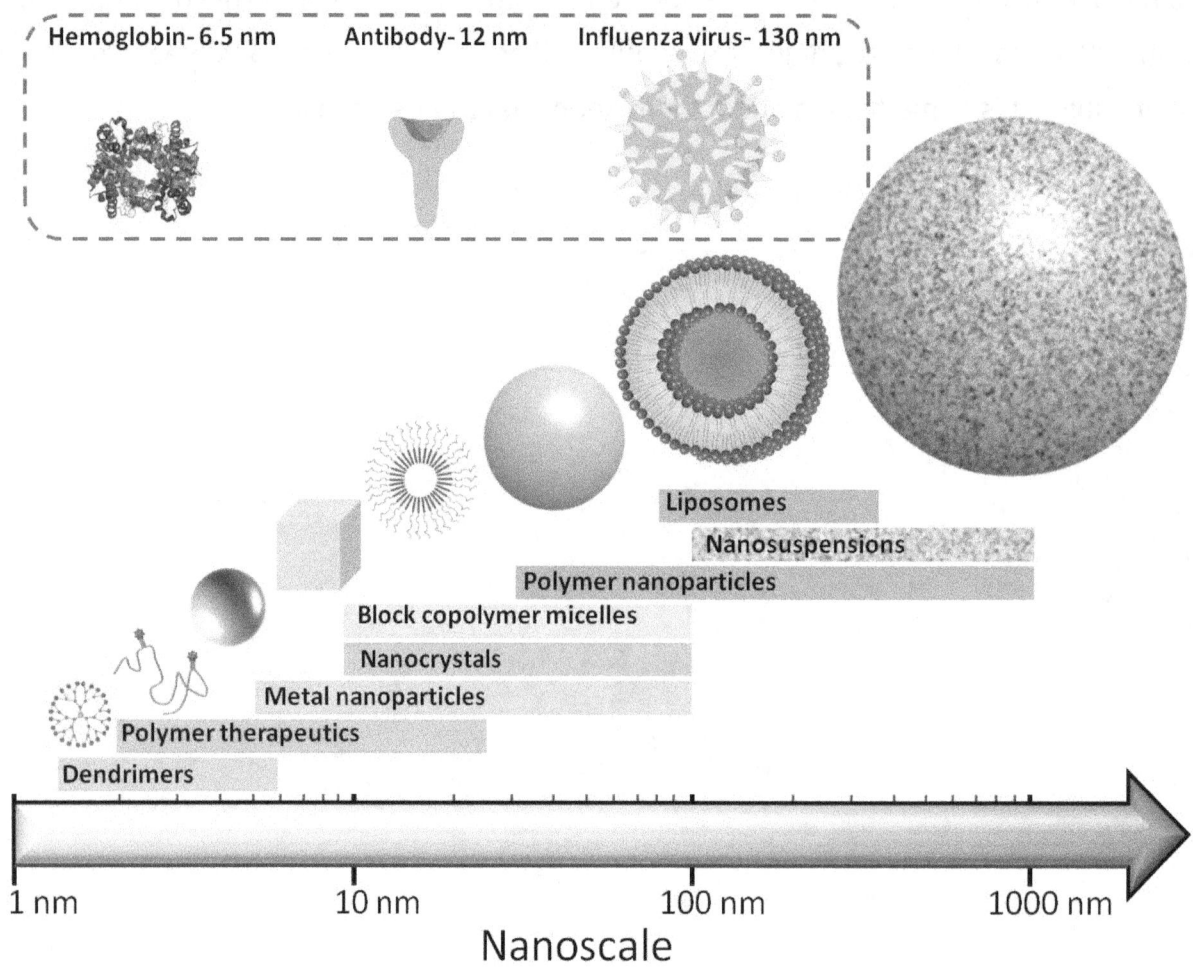

Very small shapes and design patterns can be created and inserted into the bloodstream and body to attack the bad diseases and extinguish them. This means we're taking synthetic parts and distributing them into the bloodstream to create a result of better distributed medicine.

Our bodies already have the natural ability to defend against harmful pathogens and fight colds and some diseases. With the invention of vaccines most people know that they put a little bit of the "sickness" into the body so that the body can remember it and learn to fight it, thereby preventing it from being able to destroy the body in the future to put it simply. Well, immunotherapy is a sort of vaccine which targets the immune system to put it into action to attack certain types of tumors and cancers. It doesn't involve hair loss or some of the harsher side effects like chemotherapy and radiation can cause. It is the latest way doctors are treating some cancers. It frees the restricted immune system in the body and allows it to attack the abnormal cells. It may have a chance of attacking normal cells which is a risk, but it doesn't seem to have any deadly consequences – it simply won't work in some patients. While this does not use any sort of prosthetic I wanted to include it as evidence of medicinal discoveries that can prolong life and new discoveries that are happening daily showing we're making progress.

## Chapter 12 – Life After Death?

I thought it would be interesting to dive into a topic that is sort of scary; corpses. I think one key factor in promoting longevity is studying corpses and why death happens and what happens after death. There will be answers to the questions like, "What happens to our bodies when we die?" Of course, mortuary services know many of these answers, and I have a mortician in my family I could prod for answers, but I don't think there's anything there that I cannot imagine from seeing scary movies. What changes take place when we die? What is our body asking for, besides the obvious like air and water and nutrients? When a family must decide whether to pull the plug on a brain-dead individual, is there any hope for bringing a brain back to life? I think there will be. How can we provide these connective tissues and blood and fluid samples in ways that prolongs life and revitalizes? I am afraid cremation is just giving up in my opinion. I wish there was a way to preserve the body better than embalming. And, I've imagined above ground burials where new technology can easily be administered should new developments arise. Is the morgue better than the dirt? Is the dirt better than the ashes? It's a deeply personal choice and that of the family as well. At this point, I am in favor of a proper burial.

What if we came up with solutions that can bring a body back to life? Can we save the blood and fluids? Or do they go bad? Is it too early to start to think this way? Perhaps right now, but soon I hope we can find ways to prolong life even after the body has been deceased for some time. After all there are plenty of stories out there where patients have died and been pronounced dead by the doctors and have come back to life. Joe Tiralosi of Brooklyn, NY died in a hospital after going there because he felt very sick was pronounced dead and resuscitated with medications and about 4,500 chest compressions after death, which is another way doctors are having better luck bringing people back to life.  Preventing brain damage from happening is a key factor as well, which coldness can help with. There was a 2-year-old girl who drowned in an icy stream and because of the low temperature of her body it prevented brain damage by preserving the brain while doctors figured out how to get her heart going again, and after one hour of death the doctors could bring her back to life. She lived on as normal.

We didn't always have The Stethoscope – it was invented in 1816. And, the medical profession didn't use the techniques we use today to prepare the body for burial. So, being buried alive was a big fear in those days. Edgar Allen Poe wrote a scary story that created a sensation in 1844 about being buried alive called *The Premature Burial*. And, there was no such thing as preparing the body for burial through embalming like we do today, removing the fluids and such. It was more likely that people could be buried alive. People everywhere were frightened by the possibility.

There was a man who was found kicking in his body bag at the funeral home just prior to being embalmed. His name is Walter Williams and he is from Lexington, Mississippi. He was pronounced dead in his home not having a pulse. Doctors call it a miracle and think his pace-maker stopped working temporarily and then started back up again. There are all kinds of stories of people being brought back to life. So, why then can't one imagine that it would be possible to bring a corpse back to life if it had not suffered brain damage? I think it is possible and more doctors should try to revive patients who have been pronounced dead at least at first. I'm not saying a rotting body can be brought back to life, but there may be some ways in which life can be restored even after some deaths have occurred.

I'm not suggesting keeping brain-dead patients alive all the time or making ourselves into robots or genetically reproduced versions of ourselves, because that is not the same. But, I am suggesting we create more opportunities for fast growth when it comes to longevity. I'm not being hopeful. I'm searching for results that are concrete and useful in creating healthier and longer lives for all. I understand diet and exercise are the number one ways we can live longer as of now. And, with better nutrition, shakes and vitamins than ever; making sure our bodies are nourished is a great idea.

## Chapter 13 – Innovations Galore

Some people are taking HGH – Human Growth Hormones to revitalize their body and sex drive. It is a compound produced by the pituitary gland in the human body and is available for sale in pill and injection form and can be taken routinely. There are some nasty side effects that are thought to be linked to HGH like cancer, breast growth in men and a few others that you wouldn't want. Aging Hollywood celebrities seem to like HGH to bring out muscle tone, energize themselves, revitalize their looks; the effects are associated with anti-aging qualities.

Scientists have discovered how scorpion venom from deathstalker scorpions has been proven to make brain tumors stand out by sort of lighting them up – calling it a paint for the brain. The venom sticks to the tumors and allows for better surgical removal, avoiding removing healthy brain tissue. Dr. Jim Olson helped create the tumor paint and the FDA approved its use to aid in brain surgery removing tumors. It is injected into the blood stream where it is carried up into the brain and it latches on to cancerous cells showing on a computer screen exactly where the brain tumor is. We don't know where the innovations are going to come from in nature for medical advancements, and I'm a big believer in naturally stopping aging as well as with mechanics. If you can eat healthy foods and eat certain vitamins and keep your body in shape you are more likely to live longer.

While I think general studies are important, such as earning a doctorate degree, I believe we need to see more specialists who are really diving deep into the possibilities of new areas of each part of the human body. For example, I've come up with 7 major body section specialists with sub-specialists for each miniscule part of each major section. The combined new facts along with creativity in design for the function of replicated body parts, mechanical or real (such as a kidney transplant) would be astounding for medical advancement. Just think of all the wonderful new ideas these specialists could come up with. I'm talking nano specialists. The 7 major parts of the body that I think should be broken down into important sections are the brain and skull cavity, smell and taste functions, rib and spine, hips and all that go along with hips including genitals, joint and rotator specialists, limbs, and skin. Then, each part divided into as many sub-sections as possible to specialize in.

I also believe it would be important while our bodies are full-grown, yet young enough to heal fast from surgery to establish reinforcements in areas that seem to become the most fragile as we age and even areas that could be the most vulnerable for us in case of accidents. For example, minimally invasive surgery where the skeleton is reinforced with durable covers made of very hard plastic and metal screwed into the bone with tiny stainless-steel screws so that our hips and legs would be sturdier as we age. It would prevent the need for hip replacement surgery at an older age where it may be riskier to have surgery.  Instead, we could reinforce the hips with a sort of coating, seal and large staples or screws holding the bones tightly in place, perhaps with rubber bumpers and a ball joint cover, like an I-phone case, so that when its dropped, it doesn't shatter. These sorts of covers could be rubber and go over the jugular vein, hippocampus and back of the skull, as sort of a built-in helmet in case of an accident. These reinforcements would be a sort of early-aged maintenance program for those who could afford it. And, again this would be minimally invasive in the future.

# Chapter 14 – A Youthful Attitude

Haven't you ever seen those women who look like they're 20 when they're 40? They *think* young. They would never let their looks go by the wayside. And they stay in shape. That's the kind of person I want to be and so far, I am doing alright. People tell me all the time I look very young for my age. I have a very youthful attitude. And I make sure to eat healthy and exercise. Plus, it probably didn't hurt that I lived in Hollywood, CA on and off for the past 10 years where so many people are concerned with looking their absolute best. I once spent $300 on facial cream and cleansing products while living out there. It is not unusual to spend a lot there.

I want to get to the future as fast as possible, and I find myself wondering why people aren't having anti-aging measures as a part of their daily life at a very young age. I started using wrinkle cream when I was a teenager and I think I have very nice skin now as a 34-year-old, perhaps because of it. I think it will be more normal in the future for people to get anti-aging shots and preventative procedures like the bone stabilizing procedure I'll discuss more later. It will be second-nature. While it isn't a good idea to become too obsessed with looks – as some people have had too many plastic surgeries and end up looking terrible. It is a good idea to take good care of yourself as you age and it can't hurt to remain fashionable and have a youthful outlook on life to be your best self. I am very thankful that I have a great attitude when it comes to aging, and that is to remain healthy and youthful in attitude and looks. I see older people cutting their hair all off and letting it go grey when they could color it and let it grow out a little. I think trying to look your best is important because your attitude can change your health.

I'm not against grey hair, either. Some teenagers are coloring their hair grey. I think some people look great with grey hair, but some people don't and it just makes them look old. It depends on the person and the haircut. But, what I am saying is that caring about looks is important and taking extra measures to make sure you're looking your best is nothing to be ashamed of – it should be celebrated! When you look good you feel good as they say.

Our bodies continually change as we age, or develop. And our bodies are constantly growing. We get more wrinkles on our hands and face and we can do things to help soften the skin with lotion, but some aging is inevitable. I personally think men look better with some wrinkles on their faces, making them look a little more rugged. I wonder if men think the same about women, or if there is a double-standard for youthful looks. Smile lines are nice.

My grandmother on my Dad's side is still alive and has a type of bone marrow cancer that is slow-growing, so she can be on out-patient chemotherapy. She still gardens daily and takes great care of herself. The doctor predicts that she has 5 more years to live and she is 78. She also has a broken ankle bone from when she fell in the forest, picking wild raspberries one day. She had surgery and they put screws in her bone to help hold the bone together better, and you can feel the screws through her skin – they are rounded. This surgery was done quite some time ago, and today and in the future, I hope more surgeries will be more streamlined, using flat screw heads and making them flush with the bone. Not just for appearance, but she still experiences pain from that surgery. Her ankle swells up whenever she walks a while. Doctors are becoming more precise with their surgeries and are creating more beautiful surgeries all the time. There are techniques that are becoming more standard that experts are developing right now. My grandma still remains hopeful and energetic and reads daily, with a great attitude.

I have an aunt who studies and practices energy healing. It is like Reiki, but not quite. It uses touch and study of the chakras to determine where a person is low on energy and where they are doing ok. It optimizes a person's energy field. I have had her treatments and they work wonders. Human touch is very healing and energizing, so I would recommend getting massages and getting hugs and experiencing touch whenever possible. Sex also creates healthier bodies and most doctors will tell you that having a healthy sex life is very good for you.

# Chapter 15 – Hope And Nutrition

For now, all we can do is hope that the pace of developing medical procedures are fast and efficient and well-supported. I am so very excited to hear about any new accomplishments that happen in the medical field when it comes to the development of prosthetics and I think there needs to be more developers and designers out there making things happen, which is why I'm writing this book. Keep in contact with me if you or anyone you know is thinking of becoming or already are a designer of prosthetics. My website is www.bethanyhealy.com. Making the designs only requires a pencil and paper really, but making prototypes are where the expense occurs. If more millionaires invested in designers making prototypes we would see more progress faster. Also, if we had more surgeons creating designs, we would have more functional prosthetics. I would encourage any doctor and surgeon to become designers of prosthetics and implants. It could also be very lucrative.

I think more surgeons need more places to go to for new and improved products. We are getting so good at surgery as humans, it's unbelievable. And, with more products for surgeons on the market, I think there could be better surgeries in the future. Right now, though, where would one go to ask for stabilizing surgery so that it makes it easier on them when they age? I think doctors would perhaps laugh in the faces of those who asked for it, and would say that it would cause even more pain as we age right now. And, the reason is because we don't yet have the technology to create the minimally invasive surgeries I'm talking about that would make it easier for the patient or client to heal fast and be better off. They are already developing minimally invasive spine surgery that can help with back pain and those types of techniques are going to be able to be a part of the new technological developments. I don't want to depress anyone or be too dark. Life is precious and we should be happy with our lives the way they are. But, it excites me to think that new discoveries are being made daily regarding living longer and I want to be a part of it.

I think it will be possible to have surgeons go into the heart and unblock the arteries that are clogged so fewer patients will have heart attacks. Heart disease is the number one cause of death in humans. Garlic has been associated with a healthy heart, because it has been shown to lower blood pressure and increase anti-oxidants. A good diet low in saturated fat and good exercise help one have a healthy heart. Smoking can cause heart problems along with lung problems. Second-hand smoke has been linked to these problems as well. I've never smoked and I hope people quit smoking who do, because I saw my grandpa die at 70 because he had a heart attack from smoking. He always coughed a lot and smoked no-filter cigarettes. He suffered a lot and could not quit smoking cigarettes, even after a bypass surgery and up until his premature death. Smoking is bad!

The use of magnifying glasses to do the surgeries in the future is going to be important, and computers can help with getting a closer view of what's really going on in the body so that the surgery can be cleaner and more precise, whatever it may be. It is harder for gravity to take its toll on toned muscles, so keeping healthy muscle tone is important. Too many people let themselves go, especially after retirement. They think that just sitting around watching television and never exercising is the life. Well that is a great way to become fat and overweight. And, it is very hard to lose weight the older you get, because you already have a slower metabolism so you don't need to eat as much as when you were younger. Will we see the day when legs that are frail are simply cut off and prosthetic strong legs are installed that can be controlled with a computer chip in the brain? We're already developing the technology. So many elderly die from falling and breaking a hip, which is why having hip cases put on early in life would be so beneficial, unless it becomes so much easier to mend a broken hip in the future. But, I think a preventative case could make the hip indestructible. If there was a brace on the hip that was built in and someone fell on that, it would merely break the fall and not the hip, preventing premature death from a hip break. And, leg muscles are weaker as we age, which calls for extra exercise as we age. Our bodies yearn to be worked out and call for more effort in that area. It isn't fun, but if you can take on an exercise that you like doing such as swimming versus running on the treadmill then you'll be more likely to stick to it. This is one thing I learned while being a health coach with Herbalife.

I believe people refer to the term *age* as an invalid reason. For example, they'll say that a heart condition was just due to *age*, or that skin deformations are just because of *age*. When, in reality the heart condition may be due to heredity and is considered an actual disease and some people experience more skin mutations than others at various times in their lives, making it a genetic problem and one that can be addressed. Age is no excuse for anything. Yes, some people develop more problems as time goes by and those problems can mainly exist later in life, like deterioration of the bone and other so-called age-related problems, but why then does it happen in some but not others? It's a disease, not *age*. People also use age as an excuse to not take as good of care of themselves, believing things like if they were younger they would exercise more and they should dress their *age*. So, older people cannot exercise or wear Kate Spade or have cute haircuts that are colored? Simply not true. These types of behaviors are also area-specific. For example, most people in Iowa do not care about wearing designer clothing, paying a lot for haircuts and many people are overweight whereas in California – particularly Los Angeles and Hollywood, people will pay enormous amounts on their hair and clothing. And age only increases the likelihood that someone will pay more for plastic surgeries, spa treatment and other procedures that have to do with being more youthful. People also make more money there compared to Iowa, but that's no excuse in my book. You don't need to spend a ton of money to make yourself appear more attractive. Some people just don't care. It's a matter of personal choice.

I was a vegetarian for 7 years, until I moved back home to Iowa to be close to family. There just weren't vegetarian restaurants to go to like there are in Hollywood, and my whole family eats meat and some of them have lived to be well into their 90's eating a healthy meat -filled diet. So, I became a meat eater again. But I think it is healthy to be a vegetarian as well. And quite fun to eat that way. I found many foods that I loved eating such as sweet potatoes, hummus and goji berry shakes and vegetarian soups.

It was much easier to be a vegetarian in Los Angeles where there are special restaurants and grocery stores catering to vegans and vegetarians. And there is a lifestyle there were more people are vegans and vegetarians, making it easier to choose that sort of lifestyle if desired. I tried to be a vegetarian in Iowa, in these small towns where my family is from, but it is very difficult because I got a lot of pushback from family and friends mostly wondering why I didn't eat meat and there was little to no selection at restaurants and grocery stores compared to California. But I did not become a meat-eater again before writing a book titled "The Health and The Joy of Becoming Vegetarian."

Visceral fat is a different kind of fat that is deposited in the abdomen and around the organs of the abdomen. People who are thin can have visceral fat, and it is very dangerous. It can lead to heart attacks, cancer and just look unsightly. Having a healthy diet and doing ab exercises is very important in maintaining a healthy body. Fat around the mid-section is particularly threatening to one's overall health because it is linked to disease. What you eat is hugely important when it comes to longevity. A diet rich in vitamins and nutrition from vegetables and fruits, along with protein and calcium among other nutrients is essential in creating a healthy body that will last and will heal better should you get something as scary as cancer.

Cancer treatment doctors many times have nutritionists that they work with and countless doctors are in business to tell their patients how to eat properly to be at an optimal weight. There are so many companies out there that try to teach clients how to eat properly. I've always followed the old food guide pyramid and I've never had a diet high in soda-pop. I weigh 130 and I'm 5'8" tall at a very optimal weight. And I work out regularly. I think everything should be in moderation. You don't want to overdo anything.

That includes dieting. Crash diets only work temporarily. A lifestyle of eating healthy foods like good quality low-fat meats, grains and fruits and vegetables with a low intake of desserts and junk foods, allowing yourself to splurge occasionally is a great idea. You can't always be perfect. Starving yourself of good foods is a bad idea because your body goes into starvation mode and it makes everything go out of whack and can cause long-term problems. Just eat a moderate amount of food and exercise.

The stomach can stretch and if you're used to over eating, you'll need to go through a period where you still feel hungry after eating the appropriate amount of food for a while until your stomach goes back down to normal size. Soon you'll be feeling full when you eat a moderate amount of food. And, just because it's noon doesn't mean you must eat lunch. It is ok to skip a meal from time to time if you're not feeling particularly hungry. Also, it is okay to eat a few more times throughout the day in smaller amounts if you're feeling particularly hungry. Do what feels natural to your body. There is no set schedule for eating. Your body will tell you when it is done digesting the previous meal and when it is low on energy and needs food. You'll become a keener listener the more you try too. Also, you need fat to digest vitamins, so eating a fat free diet will only make you crave more. Don't be afraid to eat fat in moderation. Especially good fats like from dairy, avocados and real butter. Unless your doctor recommends some other diet, eating these fats will only make you feel fuller, satisfy your appetite and help with the digestion of key vitamins from nutritive foods like carrots, peppers, and broccoli among many others.

I think people are going to start discovering the cellular-level nutrition that is going to give them more energy and life force. With the doctors discovering more nutritious shake powders and concentrated nutrition we will start living longer from these nutrients. I don't think one should only drink shakes though, because our bodies need to be able to digest solid foods as well. But, I do think they add a lot of value to life by offering levels of nutrition and products like from collogen, blue-green algae and spirulina that are hard to get elsewhere. And soon, we will be eating just the filet minion grown meat parts from laboratories that keep the cows alive. The way we're eating will change along with the times just like it has in the past. We can eat better today because we get more varieties of foods from around the world.

I take a strong passion for food, because it's something we do three to five times a day plus snacks. Why not enjoy it? I love trying new recipes and adding healthy things to my meals like fresh fruits and vegetables, especially organic. I strongly recommend taking cooking classes, reading about new recipes and trying them, and getting interested in healthy food, because it is worth it for your overall heath. Any doctor will tell you that.

# Chapter 16 – Editing Embryos And Stabilizers

Scientists are studying and practicing how to remove bad heart disease genes from embryos. What if terrible diseases like hypertrophic cardiomyopathy, and Huntington's disease could be removed from the embryo before the baby was even born? They have used technology called CRISPR-Cas9 to cut out the mutated heart disease gene from the embryos of 42 embryos. They have yet to do the clinical trials, but that's on the way. This creates endless possibilities for parents wanting to have healthy babies. And, I can't fathom why this would be controversial at all, when it helps make a human life healthier. I say go for it and keep studying new things like this! This is the future and we cannot let fears get in the way.

Stabilizers are something I mentioned earlier, and are in my opinion crucial to supporting key body parts as time goes on and our bodies change. That is one certain thing – our bodies will continually change as they do throughout our entire lives and keeping up with those changes will be a key factor in human health and optimal living. Disclaimer: am I trying to save the world? No. I am simply interested in drawing and design that could be helpful to surgeons needing artificial body parts. I love to create and I am in no way able to save the world, nor do I think I can. I hope my designs and book spark inspiration in others and that it is a good read. Now back to my stabilizers... There are areas of weakness as we age in some people, and there is always concern for an accident such as a slip and fall or car accident, which are common and both can be fatal. Creating cases for special parts of our bodies that are important to the overall function of our bodies is something I am interested in. With surgery becoming less invasive all the time, it will be easier and easier to apply these techniques to the body. You'll notice in my drawing below there is a C-shaped stabilizer for the head that could be used to prevent fractures and breakage of the skull. This could be made of a combination of very hard plastic and rubber layers along with metal. And, could be applied to the front of the skull as well as the sides to create a sort of built-in helmet. It would be a seamless design that would fit flush with the skull and very tightly and effectively screw into the skeletal material (bone) to create a sort of second bone, thereby stabilizing the original bone.

This could be done to the nose, eyes and teeth. Any bone could be stabilized. People are already getting artificial teeth screwed into their bones in their head. They last a lifetime. It is a better alternative than having bad teeth or using dentures.

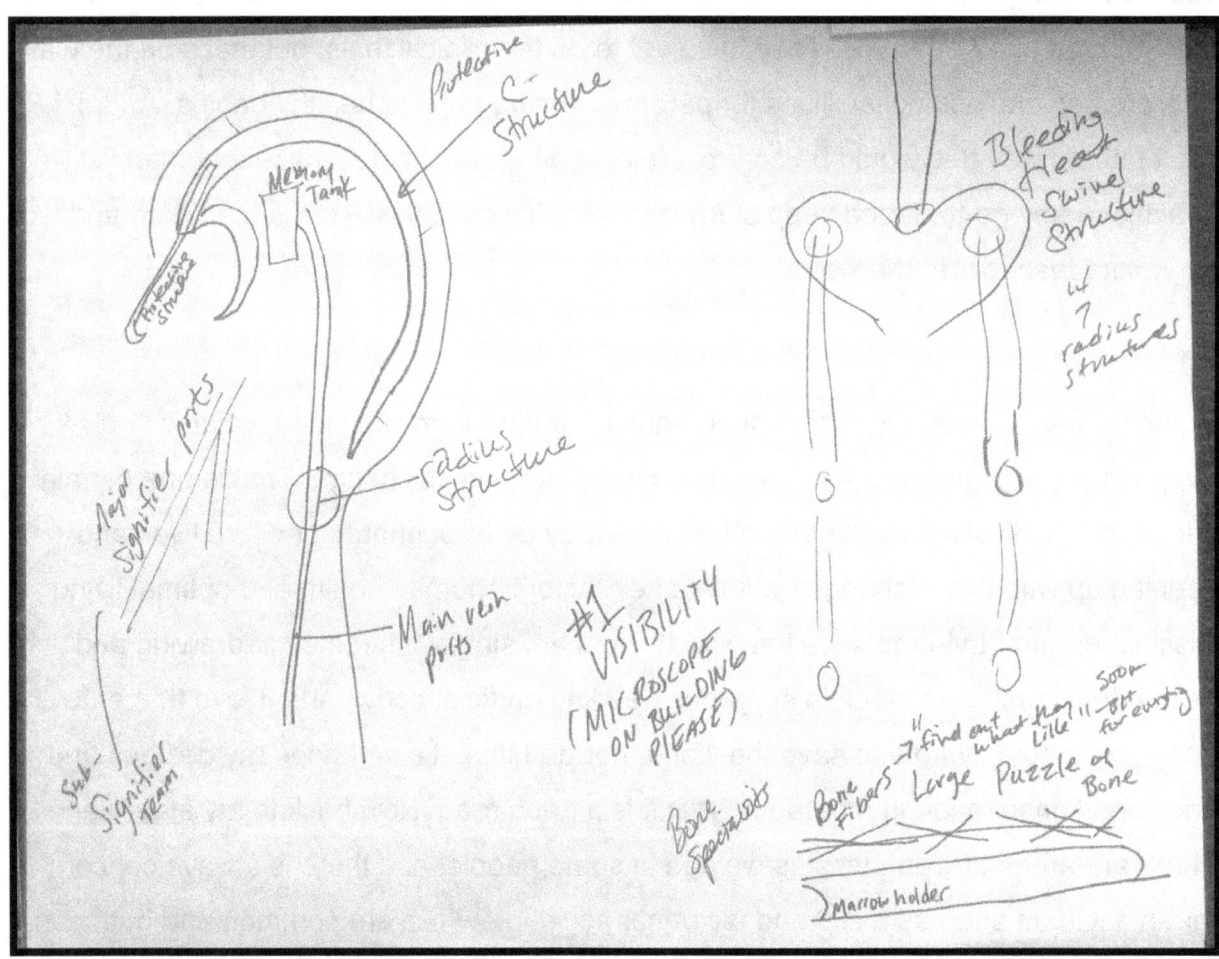

The other part of the drawing above is that the hippocampus and other memory banks within the brain could be stabilized and the jugular vein could have a second rubber-like vein around it protecting it. These protective extra layers throughout the body could be used to create a longevity of life for many who are interested. Small or large procedures could take place and people could have the stabilizers installed in a variety of ways. Joints could be stabilized or replaced, and limbs could be stabilized with stainless steel brackets or again, hard plastic – preventing fractures and the need for a cane. And, with appropriate exercise muscles could build to help stabilize the body further. So many people get lazy as they develop and our bodies are crying out for more physical movement than ever before – creating a need for more cardio. Simply taking walks and enjoying bike rides where there is a low-to-the-ground bike for safety would work wonders on leg muscles and abs. It is harder for toned muscles to sag, so it would help with overall appearance as well. Plus, there are obvious benefits to energy with exercise.

# Chapter 17 – The Artificial Kidney

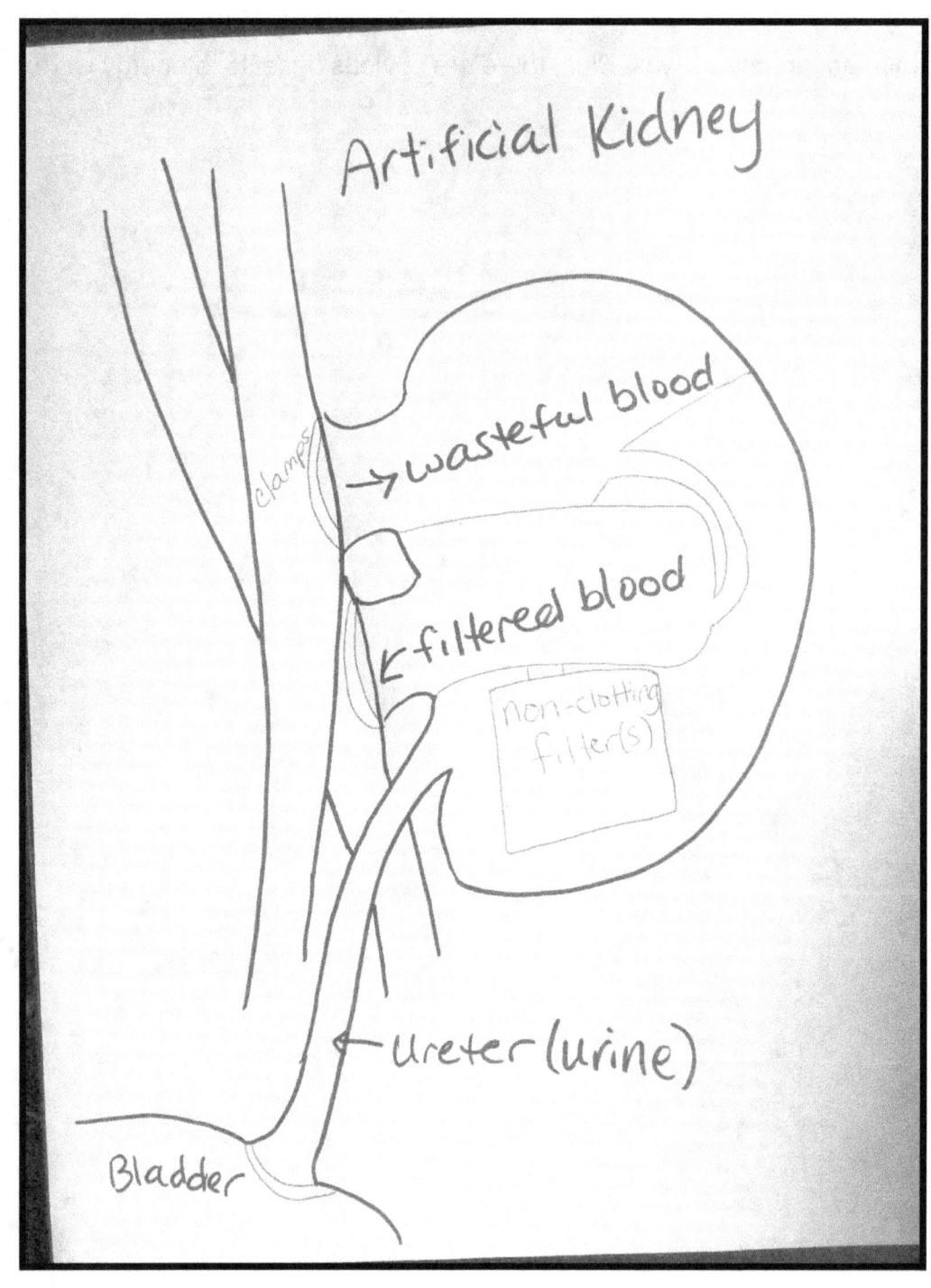

This image of an artificial kidney looks like the real thing, only is made of synthetic materials and only needs maintenance every so often. The kidney would be made of hard plastic, and covered in soft silicone in a comfortable kidney shape. It is close to the skin, near the bladder, so it is easy for surgeons to go in and make filter changes and updates. It could be USB powered and have a cord that extends out of the body to charge the motor for the filters. The filters would be non-blood clotting through a slight movement pushing blood away from the filters and a scraping off the filters periodically, like windshield wipers. It would only require a little bit of power, making it easy to charge. Also, there is a variety of ways that it could attach to the body with stitching, clamps, inner clamps that are attached to a tube, tiny Velcro-like attachments that connect to the fragile skin of the veins and bladder. And, even the possibility of a stretchy rubber material attached to an attachment on the hip bone for stability. There are countless possibilities for the design and it's just a matter of getting the right developers to make a few prototypes in the future. I will be considering this soon with grants and proposals. I will also be writing to people in the design field who are already coming up with designs to see if there is anything I can add to their work.


Where there needs to be filters that are designed to be self-flushing so they prevent blood clots for better tubes in the body we can design them. Designs within designs are important. Being a person who can think outside the box is crucial to developing great results in prosthetics and artificial valves and the like. That is why it is so crucial to blow up images. More work can be done to the design this way. Adding all the picturesque detail and fine printing can always be added later and may not even be necessary. I see kidney prosthetics being developed that look nothing like a kidney, and they are right for developing the functional product. But, I think the designs can be enhanced by making them fit in to the body better and making them true to the looks of a real kidney. I'm not complaining though. It's a start, and whatever works!

The image below is a picture of what the artificial kidney developed by the University of California San Francisco looks like. It is about the size of a soda can or a coffee cup.

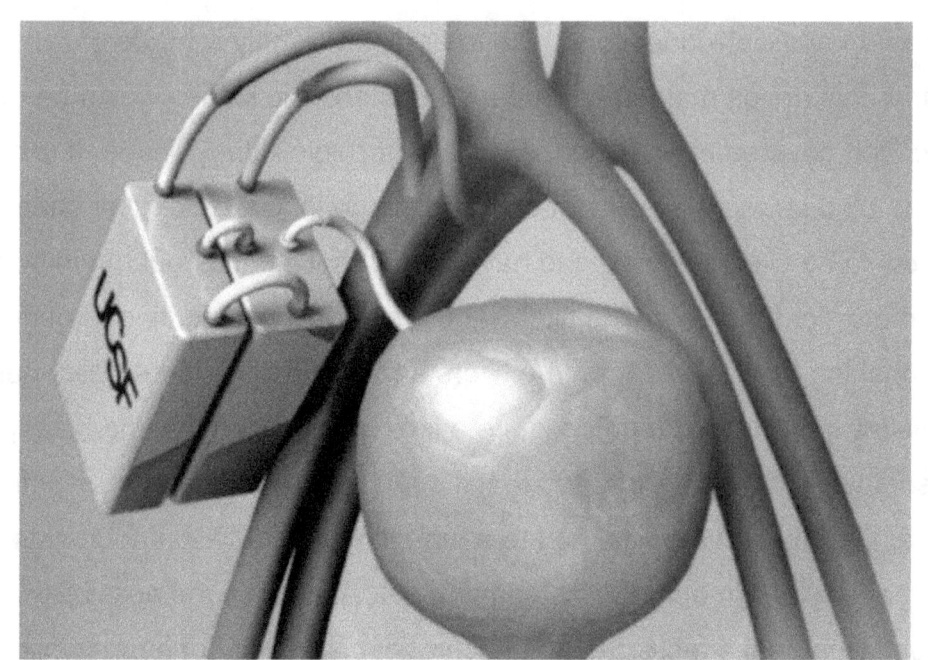

# Chapter 18 – Stem Cells

Stem cell research is another way researchers are advancing the possibility of human longevity. It's fascinating how these cells that can divide within the body to restore damaged or diseased tissue can divide and grow outside the body as well. Scientists are on their way to growing human body parts and tissues for repair and replacement for people who have diseases such as macular degeneration, spinal cord injury, stroke, burns, heart disease, diabetes, among others. Scientists are also developing meat made of a biopsy of a muscle from a cow that is grown in a laboratory. So far, the meat is completely edible and costs $250,000 for one burger, although it is said to not taste like a normal hamburger. But, they can do this and it is developing into something real. Imagine this lab meat being available at grocery stores in lobster-beef styles and vegetarians no longer needing to worry about animal cruelty the way they do now.

Stem cells are also being used to treat cancer. The doctors have two ways of inserting the stem cells in the cancer patient through the veins. One is with their own stem cells and the other is with a donor's stem cells. If it is their own stem cells then they try to make sure the stem cells are healthy and cancer free. If it is a donor's stem cells they try to make sure there is a close match like from a sister or brother, and they can get them from a donor registry as well although this is not preferred. The donor is put under anesthesia and a needle is inserted in the hip bone where the most bone marrow is concentrated and then that marrow is frozen and inserted into the patient like a blood transfusion through the vein. The stem cells then travel to the bone marrow where they help produce healthy new cells and this process of creating the new cells generally takes the body 10-20 days.

New technology is so incredible! And, it's only getting better every day. There are amazing possibilities on the horizon and they are only going to grow. The healthcare industry has been criticized for being money-driven and it is to a certain extent – very expensive in some cases. Hopefully we will move in new directions and there will be room for more competition regarding the development of new medicines and technologies, because it should be that way for the sake of everyone. We need to promote innovation more than ever right now to make leaps and strides in what's new.

Chapter 19 – Cell Regeneration

A fascinating aspect of what doctors and scientists are now discovering is the fact that we are capable of regenerating cells in the human body, according to recent studies. While we can't grow an entire limb again like a salamander can, they are studying what makes the blastema (accumulation of stem cells) form at the point of damage on the Salamander. In a study done on mice, damage from stroke and leg damage which otherwise would have caused amputation from the leg turning black, scientists have found the leg turning pink and the mouse being able to run again from a chip installed on the outside of the skin that carried cell nutrients back to the body for regenerative purposes. This new technology is so revolutionary that in the next five years, it is expected to be passed by the FDA for use on humans in healing wounds and

possibly helping heart disease and stroke victims heal. There is no need for a hospital for the application – it works remarkably well with just an application of the cell properties on the chip which can be applied and absorbed by the affected area.

David Sinclair, a Harvard Medical School Professor of Genetics, is the senior author on discovering how to keep dying cells alive in the body that cause aging. He and his team are finding that in mice, a genetic code found in the body is being discovered that creates better communication between the nucleus and the mitochondria in cells, creating properties associated with younger mice. It is said that it is comparable to a 60-year old human reverting back to a 20-year-old human in properties. Usually, as mitochondria break down as our bodies develop, or age, over time we get diseases such as Alzheimer's, diabetes, and possibly cancer as it most associated with age. This research is very new, and is still in early development stages, but results have been very promising for the correlation and use in humans.

## Chapter 20 – Mental Clarity And Direction

Mental Illness affects 1 in 5 people, and there are 3.5 million people in the U.S. living with schizophrenia. More than 21 million people worldwide are affected by schizophrenia. $175 million dollars have been granted to schizophrenia research since 1987. Schizophrenia is not what it is stigmatized to be. People do not have split personalities, and it is treatable for most people who do not have extremely severe cases with medication and behavioral therapy helps. What happens is the brain experiences delusions such as the feeling that others can hear thoughts, or that one is being watched or that others are out to harm them. People can also hear voices that aren't there, but that doesn't have to happen for someone to be diagnosed with schizophrenia and it can take on many forms. There are other disorders like bi-polar, anxiety and depression that can affect the brain, but most of these disorders can be treated with proper medication. We've come a long way in treating mental illness in just the last 30 years. And, people are understanding it more every day. We used to never talk about going to counseling or getting treated for a mental illness. But, now it is normal for people to seek help from an outside source and get help with mental issues. I wanted to include this because quality of life matters – not just longevity. With today's medicine, it is possible to change the chemicals in our brains so that we have clearer thoughts and more sound judgement. It is amazing and I wanted to share how incredible these medicines are for the future of mental health in everyone suffering from depression, bi-polar and schizophrenia among other mental illnesses.

I wanted to include a little about the history of the way people were treated. As you know, people had seriously zany ideas about how to treat conditions in the past. Lobotomy was a practice whereby doctors would stick a device up a patient's nose and make an incision into the prefrontal lobe of the brain cutting off important parts of the brain making the patient sometimes dead. 57 lobotomy deaths occurred in a Swedish state mental hospital, and the rate was 17% at its highest. It was thought to stop emotions from being so "intense". 84% of patients that lobotomies were performed on in one hospital in Sweden had schizophrenia. Patients would spend years in hospitals – sometimes 10 years, and there was a lot of overcrowding. Even today, side effects of too much medicine can be very dangerous, causing problems such as severe facial twitching among other problems such as increased risk of suicide from abnormal chemical reactions in the brain. It is important to take your health into your own hands and help yourself learn more by doing research and getting second opinions whenever getting professional advice. After all, many doctors did not get A's in their classes and barely passed their studies and still became doctors. Not all doctors are credible. Not long-ago people were held in cages with strait jackets on as "therapy."

Another medical phenomenon in history is what's known as trepanation, the oldest documented surgical procedure known to man. This is a process whereby numerous ancient skulls of healthy individuals were found with holes drilled into the back of their skulls near where a high ponytail would be. They were healthy skulls of various ages, mostly those in their early 20's and 30's. Some were as old as 50 and some were as young as 12. The reason behind all the old-fashioned surgeries is still debatable, but substantial evidence has made scientists and archaeologists believe it was from a ritual letting "spirits or demons" out of the soul and psychosurgery. It may have also been due to headaches, epilepsy, and abnormal behavior. Some of the skulls found are believed to have died from the surgery and some have healed bone matter around the hole, suggesting they lived for several years after the operation.

More than ever people are seeking programs that help with organizing their lives such as getting life coaching, business coaching, getting education, going to see a counselor and meditation. These things can lead to greater life satisfaction and can help a person become more successful. What's nice about people who are getting older is that they are continuing to do things that satisfy them in life like taking on a part time job, doing hobbies like gardening, reading and crafts, and keeping busy in general to make their lives better. Some people choose to just sit in front of the television for hours and hours and get fat and create many health problems for themselves. That is not the way to go. People think that they are doing themselves a favor by relaxing their retirement away, but really, they are making their lives shorter and the health problems like obesity and disease may cause them to be very unhappy.

It is important to do things. Take that yoga class, take that art class and don't be afraid to get out there and do things to enhance your life. This is how we make accomplishments that make us happy in life. And, when it comes to skincare, it is so important to wear sunblock to prevent age spots and skin cancer. Also, alpha hydroxy lotion products work wonders on plumping the skin and removing unwanted lines and wrinkles. There is amazing skincare product out there right now that can make us look and feel younger-looking. And many men are now getting involved in looking younger with using anti-wrinkle creams whereas before they didn't think it was "manly." But more than ever before people are wanting to appear more youthful and are not accepting "aging" which is the damage of cells, like they used to.

You can go to college at any age. There are centenarians who are still practicing being a doctor daily and going to work as a financial advisor with a family-owned business. I don't want to give false hope to anyone. It is rare to be a centenarian, but it is projected that we will have 3.7 million centenarians in the world by 2050. The USA has the most centenarians, at 2.2 centenarians per 10,000 people. You don't have to retire completely, and you can always change your mind. There are countless ways to remain active, which is what one centenarian recommended as the key factor to aging well. Some people say a little excess is okay, and there's no reason to obsess about anything. Keeping a healthy mind-set about it all is important. There is no need to become negative or obsessive about anything, and analyzing life is not only fun but a good idea to do from time to time. A Harvard study on what makes people the happiest as they age is not money or success, it is relationships – with family and friends. Stay in touch! Let's all celebrate ripening and live our lives as if every day was the last! Carpe diem!

**Credits, Research and Readings:**

https://www.google.com/amp/www.pewresearch.org/fact-tank/2016/04/21/worlds-centenarian-population-projected-to-grow-eightfold-by-2050/%3famp=1

https://www.ncbi.nlm.nih.gov/m/pubmed/17990197/

https://www.ncbi.nlm.nih.gov/pmc/articles/PMC4427816/

https://hms.harvard.edu/news/genetics/new-reversible-cause-aging-12-19-13

Eurostemcell.org

https://www.cancer.gov/about-cancer/causes-prevention/patient-prevention-overview-pdq

https://www.google.com/amp/www.cbsnews.com/amp/news/normal-weight-high-abdominal-fat-more-deadly-than-obesity/

https://www.bbrfoundation.org/research

https://www.cancer.org/treatment/treatments-and-side-effects/treatment-types/stem-cell-transplant/types-of-transplants.html

Ted Talks Episodes on Cultured Meat And What Makes People Happy As They Age

https://stemcells.nih.gov/info/basics/7.htm

http://news.nationalgeographic.com/2017/08/human-embryos-gene-editing-crispr-us-health-science/

http://jn.nutrition.org/content/136/3/736S.full

http://www.medicinae.org/e16

http://www.huffingtonpost.com/2014/02/28/walter-williams-body-bag_n_4874394.html

http://www.cbsnews.com/news/brought-back-from-the-dead/

http://abc2.org/press-blog/2014/09/fda-approves-scorpion-venom-based-tumor-paint-brain-tumor-clinical-trial

https://www.organicnewsroom.com/best-hgh-supplements/

http://blog.dana-farber.org/insight/2016/02/what-are-the-side-effects-of-immunotherapy/

http://www.britishsocietynanomedicine.org/what-is-nanomedicine.html

Hair follicle and Artificial Kidney Google Images search.

https://www.google.com/amp/s/singularityhub.com/2015/11/15/first-human-tests-of-memory-boosting-brain-implant-a-big-leap-forward/amp/

https://www.google.com/amp/s/www.technologyreview.com/s/513681/memory-implants/amp/

# About The Author, Bethany Healy

I was born in a small town called Elkader, Iowa where there are only about 1,000 residents. My grandfather was a town "Mr. Fix It" and had a shop called Bob's Repair where he worked on various items for the townspeople and was also a blacksmith, molding horseshoes and artwork out of metal. I could see the possibilities with my mother being a painter of furniture and many creative people in the family, including songwriters, crafters and artists. It has always been in my nature to be a creative person and I believe anything can be created with machinery and proper planning. It wasn't until I experienced two deaths in the family from cancer that I started thinking about how the human body could be created from machinery and robotics and became interested in prosthetic design. I have one brother and his wife and they gave me 4 nephews and 1 niece. I grew up Christian, going to Catholic church and a First Congregational church. Although I don't attend church regularly anymore, I have a strong faith in God, which I believe to be a universal wisdom and energy that is all-knowing and of a loving spirit. I have an Associate's degree in Music and I'm about 5 classes away from a Bachelor's degree in Business Administration. I am a big reader and studier of human anatomy along with keeping up with the latest news on medical advancements. I have watched many videos on the human body and am highly interested in longevity of life and human health. Please keep in touch with me on my website www.bethanyhealy.com.

## About This Book

If you've ever wondered what the future may hold knowing that we're living longer today than ever before, this book is for you. And, if you enjoy being an artist, nurse, doctor, medical professional or scientist this is a great read to inspire you. Diving into the depths of death and how we, as humans, can possibly live even longer than we already are can be frightening. But, this book gives a real-world approach to thinking about the human body and all we are capable of doing for it. It discusses the future of research that is being done and shares insights about the world and society as a futuristic mind-set while diving into the history, too. Giving hope to the hopeful and new insights to those in disbelief, this book will take you on a journey through the science of longevity of human life and all the possibilities that exist today. From the way relationships and education will exist, to the way the world will work regarding healthcare and education, this book has all the juicy forecasts that will inspire and transform your thoughts about the future. We are making strides daily, and we can do it – we can achieve longer-lasting lives! Studies from scientists and doctors are looked at as well as new ideas for designs of every human body part. Ideas are shared about the future of viewpoints of society as a whole, and mesmerizing facts are shared about what's already been accomplished to create more human longevity. Get ready to be astonished at the hope for the future of the medical world!

www.ingramcontent.com/pod-product-compliance
Lightning Source LLC
Chambersburg PA
CBHW080837170526
45158CB00009B/2580